"You cannot get through a single day without having an impact on the world around you. What you do makes a difference, and you have to decide what kind of difference you want to make."
  - Jane Goodall

We Are One: A User's Guide for Young Earthlings
Written by Dorna Schroeter
Illustrated by Kenzie Raposo

© Inspired by Nature Books 2025
   Rhinebeck, NY 12572

For teaching resources and more, visit www.inspiredbynaturebooks.com

_____

Publisher's Cataloging-in-Publication Data
Names: Schroeter, Dorna, author | Raposo, Kenzie, illustrator
Title: We Are One: A User's Guide for Young Earthlings / written by Dorna Schroeter and illustrated by Kenzie Raposo
Description: First Edition | Rhinebeck, NY: Inspired by Nature Books, 2025
Audience: Ages 6-9

Summary: A celebration of our one-of-a-kind planet and all who live here, with a look at some ecological challenges and how to choose more eco-friendly behaviors.

LCCN: 2025914667 | ISBN | 978-1-966997-04-7 (hardcover) | 978-1-966997-03-0 (paperback) | 978-1-966997-05-4 (ebook)

Subjects: LCSH: | Nature & the Natural World / Environment--JUVENILE FICTION. | Science & Nature / Environment--JUVENILE FICTION. | BISAC: JUVENILE FICTION / Science & Nature / JUVENILE FICTION / Science & Nature / Environment

Classification: LCC PZ7.1.S336548 Be 2025 | DDC [E] - dc23

Lexile: 570L

For all the children of the Earth, young and old, two legged and more, winged and finned, known and unknown, liked or unappreciated. So that all may live in a healthy and beautiful world.

WE ARE ONE

A USER'S GUIDE FOR YOUNG EARTHLINGS

WRITTEN BY
DORNA SCHROETER

ILLUSTRATED BY
KENZIE RAPOSO

Hello! I'm so glad
to see you!

I am Earth!

The third planet
from the sun.

I am your home.

I'm not like any
of my neighbors.

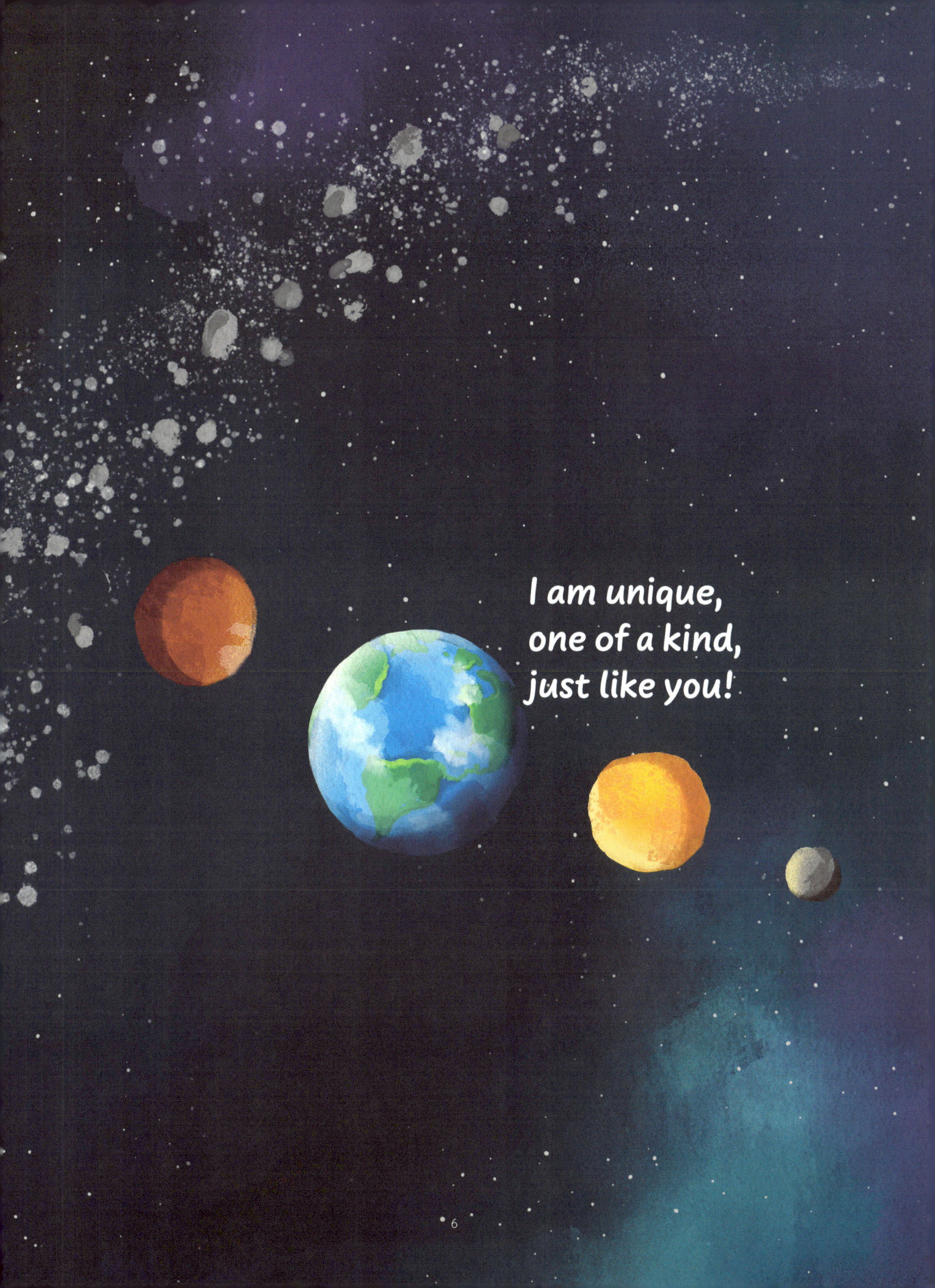

I am unique,
one of a kind,
just like you!

I am wrapped in a clear blanket of air. Fluffy clouds carry rain and snow across my boundless, blue sky.

I'm covered with...

PEACEFUL PONDS

SPARKLY STREAMS

SHALLOW SEAS

TINY

ISLANDS

DRY, BROWN, DESERTS

LUSH, GREEN FORESTS

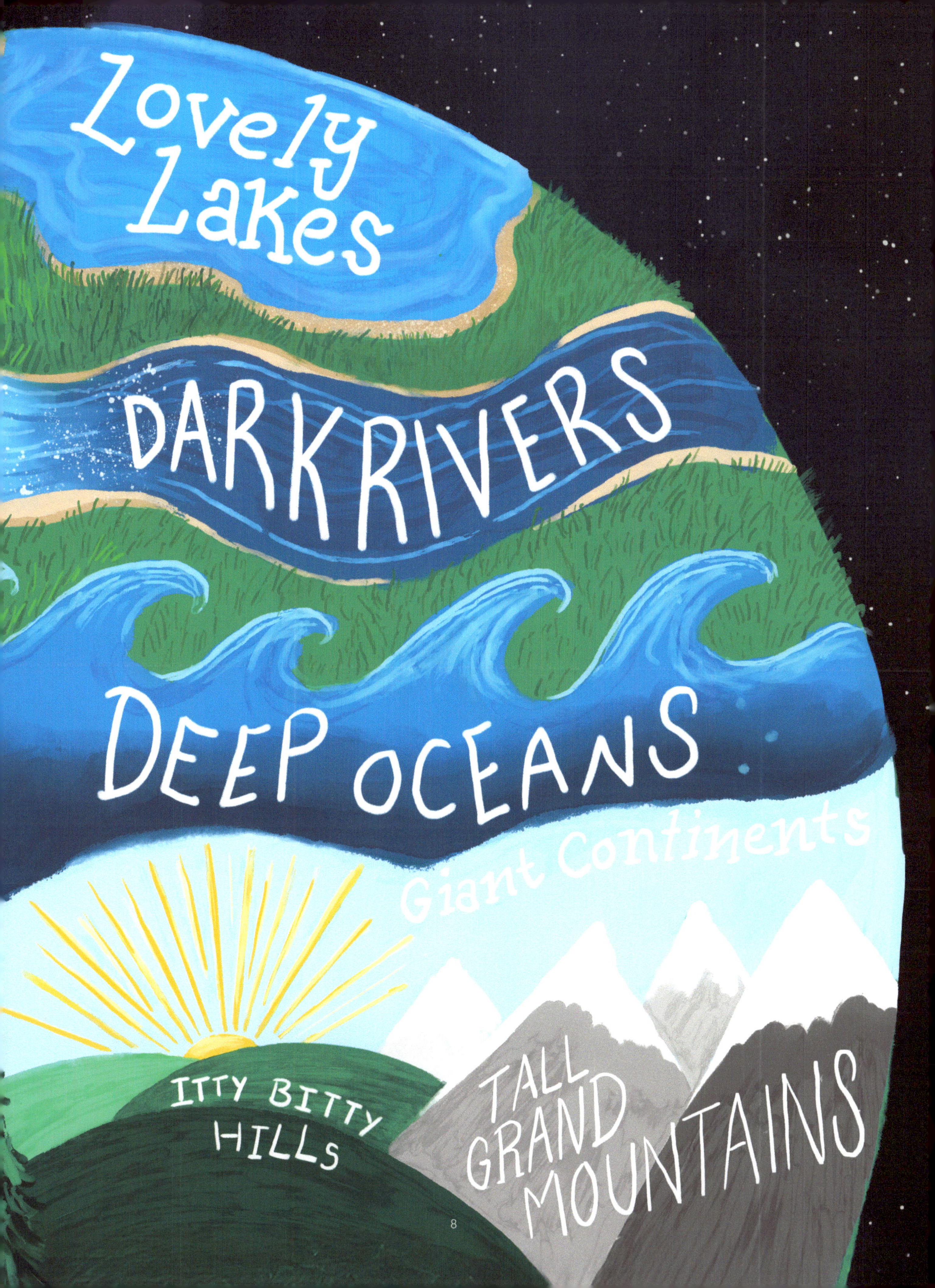

Lovely Lakes

DARK RIVERS

DEEP OCEANS

Giant Continents

ITTY BITTY HILLS

TALL GRAND MOUNTAINS

I'm beautiful!

Don't you agree?

I provide all who live on me with

water

air

soil

and minerals

Humans call these natural resources.

You use them to make many things.

Even this book!

These days, resources are being taken from me faster than I can replace them.

I may seem big,
but I'm not endless.

I can only produce so
many resources each
year for all to share.

There are many
who call me home.

Those that are
big...

and those
that are
small.

Some with two legs.

Others with four,

six, or even more.

Those with fins, and
others with wings.

Can you
think of
others?

16

Many are
known to you.

Others have
yet to be
discovered.

Some you may like...

Others, maybe
not so much!

Can you
think of
others?

All are part of one
big family that...

...depends on me
and each other.

Some ...

MAKE OXYGEN

FILTER & CLEAN WATER

POLLINATE

Others...

PRODUCE FOOD

GROW MEDICINE

RECYCLE WASTE

STORE CARBON

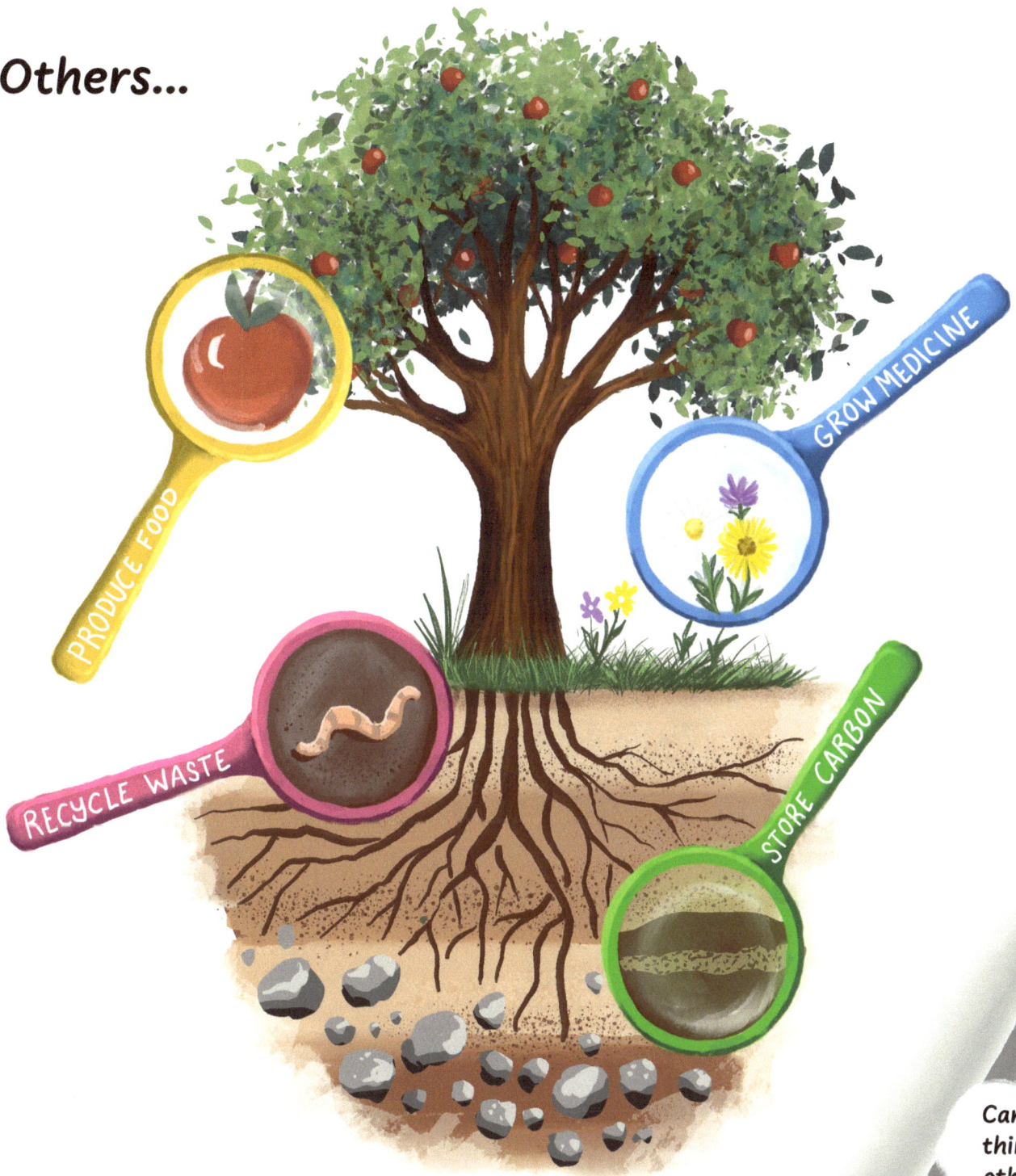

These are natural services that make life possible!

Can you think of others?

22

Lately there's waste on me that can't be recycled or composted. It's garbage blowing in my air, floating in my waters, and buried in my soil.

Even when you don't see it, the garbage is still here because **there is no place on me where things magically disappear.**

There are lots of things
you can do to help me!

Think of things you do
and use every day.

How can you do
them differently?

With each behavior change,
fewer resources are used and
less garbage is produced...

COMPOST

RECYCLE

making me cleaner, and all
who live on me healthier,
and happy again!

Start simple!

Do one change!

Then another.

And another.

Small changes add up and
over time become big changes.

Share my story with your family and friends.

With everyone working together,
we can keep the one-and-only me
and the one-and-only you healthy.

Because...

we are one

31

ISN'T EARTH AMAZING?

It's a beautiful place to live, with everything we need. We know of no other planet like Earth. That's why we must take care of our one and only home. To help us do that Earth told us:

Resources can run out if we use too much.

Plants, animals, and people all need each other.

Trash never really goes away. It always ends up somewhere.

Earth is telling us that using fewer resources and making less trash will help care for her. You will find ways to help do that on the next few pages.

Everyone, young and old, can make these changes! Soon, they will become habits you do without thinking.

Together, we can make our one-and-only planet healthier for all who live here.

USE LESS WATER

All living things need water to live. It may seem like Earth has a lot of water. But most of it is too salty to drink.

You can help by not wasting water. Here are some ways to do that.

Turning the water off when brushing your teeth can save a gallon a day!

Taking shorter showers saves lots of water. It also uses less electricity. That helps keep the air cleaner and saves your family money as well.

USE LESS ELECTRICITY

Making electricity uses a lot of resources and makes our air dirty. We call this air pollution.

You can help by using less electricity. Here are some ways to do that.

Turn off the lights when you leave a room and when you leave your house.

Turn off TVs and other electronics when you're not using them.

Unplug phone chargers when not using them. They still use energy even if the phone isn't plugged in.

To learn more, read *"Electricity"* on page 39.

USE LESS PACKAGING

Packaging holds things together. It can be made from cardboard, glass, plastic, aluminum or polystyrene foam such as Styrofoam.

Glass, cardboard and aluminium are easy to recycle. Plastics and Styrofoam are not. They stay around for a very long time and often cause harm to animals.

To use less packaging, you can:

Choose fresh fruit instead of canned or packaged fruit. It's healthier too!

Choose snacks that come in **bulk packaging.** Bulk packaging holds a lot more food with less packaging.

Choose food that comes in **cardboard or glass,** which are recyclable.

Buy **toothpaste in tablet form.** Tablets come in less packaging, and it is usually recyclable.

To learn more, read *"The Problem with Plastics" on page 40.*

PACK A NO-TRASH LUNCH

Look in the garbage cans at the end of lunch at school. They are filled with packaging, plastic bags, forks, spoons, and straws. Each of these was made to be used one time, then thrown away. That's a lot of resources and waste!

To make less trash, you can:

Pack your sandwich and snacks in **reusable containers.**

Use silverware instead of plasticware.

Use a **reusable drinking bottle.**

Do you need that plastic straw? If you do, use a reusable one.

USE LESS DISPOSABLE THINGS

Disposable means we use it once and throw it away. Things like plastic straws, bags, bottles, forks, spoons, knives, plates, and cups. Disposables make our lives easy, but **they use lots of resources and make lots of trash.** Some animals think disposables are food and eat them. This can make the animals sick and may kill them.

To use fewer disposable things, you can:

Use reusable plates, cups, and bottles.

Use **reusable shopping and produce bags.**

When eating out, bring a **reusable container for leftovers.**

Choose a **bamboo toothbrush.** Bamboo handles break down and turn back into soil. Or buy a **toothbrush with a replaceable head.**

To learn more, read *"Where Our Garbage Goes"* **on page 40.**

MAKE LESS TRASH WITH YOUR CLOTHES

Most of our clothing is made to be thrown away after a few uses. Clothes made this way use a lot of resources and make a lot of waste.

To throw away less clothing, you can:

Buy just a few pieces of clothing made from organic cotton, hemp, or bamboo. Then mix them up so you have a new look every day. These materials reduce waste and take care of our planet.

Shop at thrift stores! They have lots of great, gently used clothing.

MAKE LESS FOOD WASTE

Each year, Americans throw away about 225 pounds of food. That is about how much a baby elephant weighs! That's a lot of wasted resources and food!

Most food waste goes to a dump. There, it rots and makes methane, a heat-trapping gas. Methane leaks into our atmosphere, making it even warmer, which is bad for people, animals, and plants.

To make less food waste, you can:

Take smaller portions. You can always take more if you are still hungry!

Donate unused, packaged food to a food pantry.

Compost food waste instead of throwing it in the garbage. Composting turns food waste into soil. You can make a compost pile in your backyard or use a compost bin. Some towns even have special composting programs.

To learn more, read *"Climate Change"* on page 39.

A BETTER WAY TO CLEAN UP

Paper towels, toilet paper, and tissue are made from wood which comes from trees.

Every year, people cut down 110 million trees just to make those things! That's a LOT of trees and a lot of waste!

Trees are super important! They make the air we breathe. They also clean our air, give off water, make shade, give animals homes and food, and keep the soil in place.

Use old rags or reusable cloths. They can be washed and reused.

Buy **paper towels, toilet paper, and tissue** that aren't made from trees. There are many choices. Some are made from recycled paper. Others, bamboo, a grass that grows very fast. And some from sugarcane.

GET AROUND IN A CLEANER WAY

Cars, trucks, and planes help us get around quickly. Most run on gasoline. Burning gas pollutes our air. It also makes carbon dioxide, a heat-trapping gas. This makes the Earth even warmer. That's not good for people, animals, and plants. The good news is cars and trucks are being made to use less gas and run on clean energy.

To use less gasoline and make less pollution, you can:

Walk or ride your bike when you can! Besides not making pollution, it is great exercise!

Share a ride with someone. Using one car instead of two uses less gas.

Take the bus to school. One bus filled with kids means fewer cars on the road, less gas used, and less pollution.

To learn more, read *"Climate Change"* on page 39.

CELEBRATE WITHOUT HARM

Balloons are bright and colorful. Letting them go is fun and pretty, but they always come back down. When they do, they stay on the ground for a very long time. Animals can get stuck in the strings or think the balloons are food. This can hurt or even kill them.

To do less harm when you celebrate, you can:

Make bubbles!

Create and release seed bombs. Make them by mixing together clay, compost, native wildflower or tree seeds and a little water. Roll into balls.

Fly kites!

ELECTRICITY

Our lights, TVs, phones, and computers need electricity to run. Electricity also helps us cook, keep food cold, and make hot water.

A lot of our electricity is made at faraway power plants. Long wires, called power lines, carry the electricity to our homes, schools, and other buildings.

To make electricity, power plants burn coal, gas, or oil. Burning these makes dirty air called pollution. It also makes carbon dioxide.

Too much carbon dioxide is bad for people, animals, and plants. To learn why, read about climate change in the next box.

The good news is that more people are using clean energy made from the sun, wind, and moving water. The tricky part is that the sun doesn't always shine, and the wind doesn't always blow. But smart people are working on ways to solve this and coming up with many cool new ideas!

CLIMATE CHANGE

Around the Earth is a thick layer of gases called the atmosphere. Think of it as a blanket that prevents the Earth from getting too cold or too hot. Burning fossil fuels such as oil, gas, and coal puts a lot of carbon dioxide into the atmosphere. Too much of this gas is like adding another thick blanket. It is making our atmosphere warmer and changing our climate.

Two other things that are changing our climate are:

Cutting Down Our Forests
Trees take carbon dioxide from the air and put the oxygen we breathe back into it. When we cut forests down, there are fewer trees to do this important work.

Food In Landfills
When food waste rots in landfills, methane is created. This gas leaks out of landfills and into our atmosphere. Methane is more powerful than carbon dioxide. It's like adding even thicker blankets to our bed.

A warmer atmosphere changes our planet in many ways.

- It is getting hotter.
- There is more rain in some areas and less rain in others.
- Oceans are getting warmer, and sea ice is melting.
- Storms are getting stronger.

Each of these causes harm to plants, animals, and humans.

WHERE OUR GARBAGE GOES

Some garbage goes into landfills. Landfills are giant holes in the ground filled with garbage. After a landfill is full and covered, some garbage rots. This creates methane, a gas that makes the Earth hotter. Landfills also make a slimy liquid called leachate (sounds like: LEE-chayt). This liquid can leak into the ground and harm our drinking water.

Some garbage is burned to make electricity. That sounds helpful, but burning trash makes dirty smoke that pollutes our air.

Some garbage is dumped into our oceans. Just because we don't see it doesn't mean it's gone. Sea animals like turtles, fish, and whales can be harmed by it.

Some people think we should put garbage into rockets and send it into space. This will only create other problems. Blasting rockets into space makes a lot of pollution. Floating trash could crash into satellites (the machines that help us talk on phones and watch TV). Plus, it would cost way too much money!

THE PROBLEM WITH PLASTICS

We make plastic from oil. This is a natural resource that will run out someday. Every year, the amount of plastic we make is equal to the weight of 30 million elephants. If all those elephants stood in a line, they could wrap around the Earth over 5 times!

Many plastic things, such as bottles, bags, straws, and snack wrappers, are made to be used only once, then thrown away.

A lot of plastic ends up in the ocean. Each year, it almost equals the weight of 3 million elephants! Some animals get tangled in it, and can drown.

In the water, plastic breaks into tiny pieces, like glitter. These tiny bits are called microplastics. They are everywhere! Many sea animals, like fish and turtles, eat them. Over time, the tiny plastic fills their stomachs. This makes them feel full, so they don't eat and starve to death. It can also build up in their gills, making it hard for them to breathe.

Some plastics are buried in landfills. Over time, they break down and make harmful chemicals that leak into the ground.

Other plastics are burned. Burning plastic pollutes our air. That can hurt people, animals, and plants.

FOR THE TEACHERS, LIBRARIANS & PARENTS

Few people know the basic operating principles that keep our planet functioning properly. *We Are One* focuses on three principles: *We are all connected, There is no such place as away,* and *The Earth has limits.*

Learning how our planet works helps us better understand our one-of-a-kind home. This helps us to see how our daily choices often don't align with what our planet needs to stay healthy. When Earth is healthy, its systems and living things, including humans, are healthy.

Here's more information about the three principles introduced in this story.

We Are All Connected and Depend on Each Other.

All living things are part of a big interconnected web. Humans are just one species among the approximately 8.7 million in this web. This includes 6.5 million on land and 2.2 million in the oceans. Each has a right to exist, and each plays an important role in its ecosystem. We refer to what they do as natural services. These include:

- Oxygen production
- Water filtration & purification
- Air purification
- Nutrient cycling (decomposition)
- Food production
- Soil formation
- Pollination
- Weather protection
- Temperature regulation
- UV protection
- Pest control

All living things depend on these natural services to survive. But when people burn, build, mine, pollute, or use too many resources, they hurt and kill the organisms that provide these services. This threatens all life, including humans.

The health of one part of the web of life affects the health of every other part. Put another way, if one part of the web is harmed, the entire web suffers harm.

Most current human systems and behaviors are based on the belief that people are separate from nature.

There is No Such Place as Away. Everything Must Go Somewhere.

Nature wastes nothing. Everything gets reabsorbed back into the system and reused again and again. When plants die, they break down and become nutrients for other organisms. In the oxygen cycle, animals breathe in oxygen released by plants during the photosynthesis process. Plants breathe in the carbon dioxide produced by animals during their respiration process.

Humans consider the production of unusable waste, whether in our homes or on the industrial scale, normal. Although some waste is recyclable, most isn't. A lot of it is harmful. We are the only species that produces waste that can't be reabsorbed back into the system.

We're running out of places to bury our waste. Burning it doesn't make it disappear. It just turns into ash and gases that all living things breathe. Dumping it into the ocean doesn't solve the problem either. It just moves it somewhere else, where it harms the animals and plants living there.

Most current human systems and behaviors are based on the belief that there is a magical place called away.

The Earth Has Limits.

Even though Earth looks big, it doesn't have endless resources. Everything we need, including water, forests, fuels, minerals, plants, animals, and air, has limits and must be shared by all.

To get these resources, people dig, drain, drill, kill, and cut down. Right now, humans are using more resources in a year than the Earth can replace. In fact, if everyone lived like most Americans, we would need the resources and space of five Earths to sustain us!

Humans are smart and creative, and our inventions have made life safer and more comfortable. But our thinking has also made many people believe the Earth has no limits and they can take as many resources as they want.

Can we heal Earth by ourselves? No, but small changes by each person make a difference and add up. When multiplied by millions, they can have a huge effect. Together we can change how we as a society think and act. It starts with one person. In introducing this information early, parents and educators can set the stage for lifelong habits that benefit all members of the web of life.

For further information:
- *The Closing Circle* by Barry Commoner
- *Response-Able How to Live Well Over Time on Planet Earth* by Jaimie Cloud
- *This Spaceship Earth* by David Houle & Tim Rumage

Use these questions as a guide before buying something.

- **Is it a NEED** (something you need to survive, i.e., food, shelter)? Or is it a WANT (something you would like but don't need to survive)? Is it just something cool that you want? Do you want it because everyone else has it? Do you want it because companies and social media say you need it?

- **Where did it come from?** Is it made from renewable or non-renewable resources? Renewable resources can be used over and over. Nature replaces them naturally. Some examples are oxygen, fresh water, wind, soil, and solar energy.

- **Maybe you know someone who has one you can borrow.** Or maybe they don't use it anymore. Could you trade it for something you don't use?

- **Can you find a gently used one** instead?

- **Is it made from recycled materials** or was it made using new resources taken from the Earth?

- **If it's not made from recycled materials, is there an alternative brand that is?**

- **If it's broken, can someone fix it** instead of throwing it away?

- **Does it have the word "disposable" or "reusable" in its name?**

- **Where is it going when it can't be used anymore?**

Another guide is the 8 R's:
- Rethink your choices
- Refuse single use
- Reduce consumption
- Reuse everything
- Repurpose/refurbish old stuff
- Repair before you replace
- Rot/compost all organic material
- Recycle as a last option

Finally, there is Pete Seeger's suggestion:
"If it can't be reduced, reused, repaired, rebuilt, refurbished, refinished, resold, recycled, or composted, then it should be restricted, redesigned, or removed from production."

There is also good news!

We are in a period of profound innovation. Today's drivers differ from those of a century ago. Then, the rise of electricity and fossil fuels led to many innovations we now take for granted. Today, limits on resources, pollution, and rising waste are pushing innovators of all ages to rethink how we do things. Following are a few examples.

- A biodegradable plastic that dissolves in the sea within hours.
- Plastic made entirely from leaves.
- Leather made entirely from leftover fruit.
- Nearly 100% recovery of lithium and graphite from used batteries.
- A magnetic liquid system that extracts microplastics from water.
- A molten salt battery that can potentially power 100,000 homes with high efficiency.
- Noise reduction barriers made from discarded tires are keeping 300 million tires out of landfills.
- A white paint that reflects 98% of sunlight and could eliminate the need for air conditioning.

Not all of these will make it to the market. However, the shift toward using Earth's operating principles as a guide will continue to grow. We have no choice.

Innovations Inspired by Nature

Biomimicry is innovation inspired by nature. It is an exciting new approach to innovation. Bio means "life," and mimic means "to imitate." From nature, we are learning more efficient and sustainable ways to do things.

When faced with a problem, biomimics ask, "How would nature do it?" They look for organisms that have already solved that problem. They then copy that organism's structures or strategies in their design.

Following are just a few of the incredible innovations inspired by nature:

- Velcro
- Vaccine storage
- Self-cleaning paint surfaces
- Windows that birds avoid
- Better shoe grips
- Buildings that don't need heat or air conditioning systems
- Super strong surgical glue
- Plastic alternatives

Janine Benyus introduced the world to biomimicry in the late 1990s. She tells us:

"Life has been doing design experiments for 3.8 billion years. What didn't work is no longer here. What we see today are the ones that worked. Nature's best ideas. And they did it all within our planet's operating conditions."

Biomimicry will be a part of almost every job in the coming years. Steve Jobs, founder of Apple, predicted:

"I think the biggest innovations of the 21st century will be at the intersection of biology and technology. A new era is beginning."

Read more about biomimicry:
- https://sciencetrek.org/topics/biomimicry
- https://asknature.org
- https://www.learnbiomimicry.com/blog/best-biomimicry-examples

Dorna Schroeter was coordinator of the Putnam Northern Westchester BOCES Center for Environmental Education for 38 years. The program served some 40,000 students each year through a wide variety of programs. Her favorite was the marine ecology program she ran several times a year in the Florida Keys.

Dorna is also the author of *Because of a Storm*, and two books about biomimicry in the series *How an Idea From Nature Changed Our World*. She has also authored several articles in scientific and professional publications. When she isn't writing children's stories, teaching biomimicry, mentoring youth and gardening, Dorna can be found biking the country roads around her home in Rhinebeck, NY, and on multi-day trips on one of the many rail trails around the country.

Kenzie Raposo is a children's book illustrator and greeting card designer based in Lake Arrowhead, California. She draws inspiration from the forests, lakes, and changing seasons around her A-frame home. With over a decade of experience in publishing and product design, Kenzie has created artwork for picture books, greeting cards, and gifts. When she's not illustrating, she spends time with her family, paddle boarding, snowboarding, and making latte art in her kitchen.

ACKNOWLEDGEMENTS

Writing a book starts with one person and an idea but takes a village of friends and colleagues to bring it to completion. I am blessed to have an support amazing team. This story was 10 years in the making, and many were part of the process. Thanks to them, a 2000+ word outline for one of my educational programs grew into this book, which Kenzie so beautifully illustrated.

Mary Burns
Diane Madaio
SaDonna Heathman
Allison Chawla
Joanne Gelb

Dan & Robin Shornstein
Karen Karpan
Marina Lathouraki
Mary Ellis
Kerri Karvetski

Lily Ross
Jaimie Cloud
Elizabeth Bates
Cat Leist

Members of the fabulous Likes' Publishing Critique Group

Thank you!

www.ingramcontent.com/pod-product-compliance
Lightning Source LLC
Chambersburg PA
CBHW061142030426

42335CB00002B/73